MONSTER HUNTERS

FINDING SASQUATCH

By Bert Wilberforce

Gareth Stevens PUBLISHING

Please visit our website, www.garethstevens.com. For a free color catalog of all our high-quality books, call toll free 1-800-542-2595 or fax 1-877-542-2596.

Library of Congress Cataloging-in-Publication Data
Names: Wilberforce, Bert, author.
Title: Finding Sasquatch / Bert Wilberforce.
Description: Buffalo, NY : Gareth Stevens Publishing, 2026. | Series: Monster hunters | Includes index.
Identifiers: LCCN 2024050146 (print) | LCCN 2024050147 (ebook) | ISBN 9781482472110 (library binding) | ISBN 9781482472103 (paperback) | ISBN 9781482472127 (ebook)
Subjects: LCSH: Sasquatch–Juvenile literature.
Classification: LCC QL89.2.S2 W54 2026 (print) | LCC QL89.2.S2 (ebook) | DDC 001.944–dc23/eng/20241202
LC record available at https://lccn.loc.gov/2024050146
LC ebook record available at https://lccn.loc.gov/2024050147

Published in 2026 by
Gareth Stevens Publishing
2544 Clinton Street
Buffalo, NY 14224

Designer: Andrea Davison-Bartolotta
Editor: Therese Shea

Photo credits: Cover, pp. 1, 7, 15 (sasquatches), 25 Shutterstock AI/Shutterstock.com; cover, p. 1, series art (glowing compass rose) LoveDesignShop/Shutterstock.com; series art (cardinal compass rose) Angela Ksen/Shutterstock.com; series art (ordinal compass rose) 32 pixels/Shutterstock.com; series art (fact box rope border) LanaN/Shutterstock.com; series art (fact box wood) EE SPACE/Shutterstock.com; series art (map background) Olha Andreichyn/Shutterstock.com; series art (curled paper background) MIGUEL G. SAAVEDRA/Shutterstock.com; series art (flat paper) Paladin12/Shutterstock.com; p. 5 (bottom) ForeverLee/Shutterstock.com; p. 5 (top) Wirestock Creators/Shutterstock.com; p. 7 (map) Pyty/Shutterstock.com; p. 9 (background) Fractal7/Shutterstock.com; pp. 9 (map), 11 (map), 19 (map) Rainer Lesniewski/Shutterstock.com; p. 11 (bigfoot) KS-Design/Shutterstock.com; p. 13 (bottom) Edmund Lowe Photography/Shutterstock.com; p. 13 (map) Globe Turner/Shutterstock.com; p. 15 (blue map background) FrankRamspott/iStockphoto.com; p. 15 (B.C. map) Mlyons/iStockphoto.com; p. 17 (bottom) JeffHollett/Flickr; p. 17 (map) Peter Hermes Furian/Shutterstock.com; p. 19 (inset) File:Patterson Gimlin Bigfoot.jpg/Wikimedia Commons; p. 19 (film frame) Thomas Bethge/Shutterstock.com; p. 21 Aaron Green Photography/Shutterstock.com; p. 22 GSMillerimages/Shutterstock.com; p. 23 (map) File:Bigfoot-Trail-Route.jpg/Wikimedia Commons; p. 23 (bottom) Fernando Tatay/Shutterstock.com; p. 26 File:Bigfoot trap.jpg/Wikimedia Commons; p. 27 Kellyriane/Shutterstock.com; pp. 28-29 (map) Janos Levente/Shutterstock.com; pp. 28-29 (paper) AKaiser/Shutterstock.com.

Printed in the United States of America

CPSIA compliance information: Batch #CSGS26: For further information contact Gareth Stevens at 1-800-542-2595.

CONTENTS

Words in the glossary appear in **bold** type the first time they are used in the text.

SASQUATCH HUNT!

What was that? Did you see a large, brown, hairy "thing" disappear into the trees? Sure, it could have been a bear . . . with really long arms . . . that walked upright like a person . . . . Are we thinking the same thing? That might have been Sasquatch!

No one has ever proven Sasquatch, or Bigfoot, is real. But plenty of people believe. In this book, you'll track Sasquatch stories using maps. You might be the first to find proof that will make everyone believe.

Sasquatch is often reported to look like a primate, which is the **mammal** group that includes monkeys, apes, and humans.

Behind the Myth

The name "Bigfoot" comes from giant footprints found in northern California in 1958. These were likely faked, though!

NORTHERN NORTH AMERICA

Most Sasquatch sightings take place on the **continent** of North America. The map on the next page shows North America's three largest countries: Canada, the United States, and Mexico.

The map also includes a compass rose, a tool that points to directions on the map. This compass rose has the cardinal directions: north (N), south (S), east (E), and west (W). It also has the ordinal, or intercardinal, directions. They are northeast (NE), southeast (SE), southwest (SW), and northwest (NW). These are exactly between the cardinal directions.

The northwestern United States and western Canada have the most Sasquatch sightings. Use the compass rose to locate these areas.

THE PACIFIC NORTHWEST

Head first to the northwestern United States. Sometimes it's called the Pacific Northwest. The Pacific Ocean is to its west. Most people include the states of Washington, Oregon, and western Idaho in this area. Some include northern California and southern Canada too.

There are big cities in the Pacific Northwest, such as Seattle, Washington, and Portland, Oregon. It's also famous for its beautiful **geographic** landforms, including mountains, **volcanoes**, and deserts. It has many forests too. And forests are a perfect place for shy animals—and maybe Sasquatch?—to live.

Use the compass rose to find Olympic National Park in western Washington.

Part of Olympic National Park is a temperate rainforest. It gets cooler weather than a tropical rainforest does, but still gets more than 100 inches (254 cm) of rain a year.

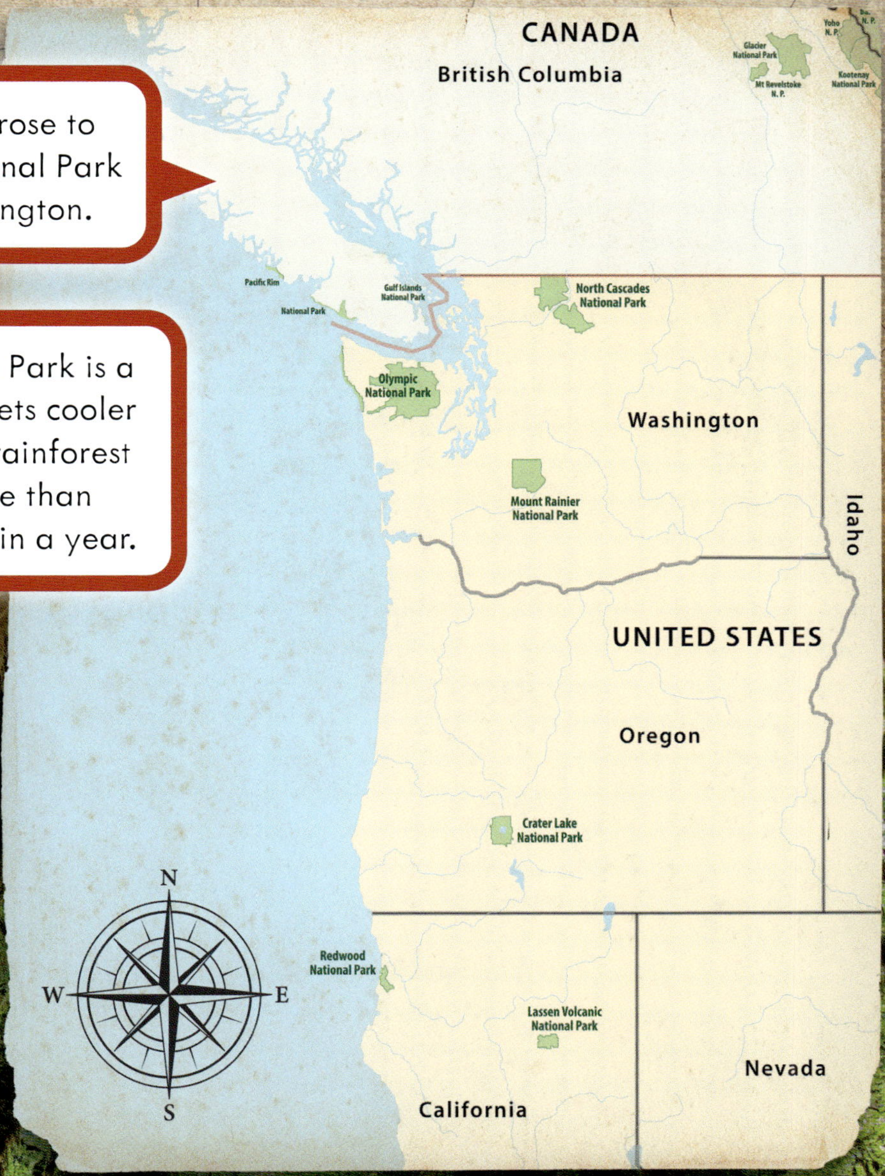

WHAT THE SALISH SAY

Salish **Indigenous** peoples have lived in the Pacific Northwest for more than 10,000 years. Salish is the name for the group of languages they speak. The name "Sasquatch" comes from a Salish word, *se'sxac* (sas-kwahch). That means "wild men."

Several Indigenous peoples live on Washington's Olympic **Peninsula**. They have many tales of large humanlike beasts living in the forests. In one old story from the Klallam peoples, a hairy creature came out of the forest and warned them a terrible flood was coming. The creature saved their lives.

The Lower Elwha Klallam Tribe lives on the north coast of the Olympic Peninsula, west of Port Angeles, Washington. Use the compass rose to check these directions.

Behind the Myth

Scientists discovered proof that huge waves of water called tsunamis flooded the Olympic Peninsula coast long ago. The Sasquatch story was a way to explain how people survived.

Not all Salish Sasquatch stories are from long ago. Head to the Lummi **Reservation** next. It's northeast of the Olympic Peninsula. Check the map to make sure you're going in the right direction. In just the 1970s, there were more than 100 Sasquatch sightings on Lummi land!

The Lummi Reservation is within Whatcom County, Washington. In 1992, the county government announced the area was a "Sasquatch Protection and **Refuge** Area." Be sure to take a Bigfoot tour while you're there!

You could sail a boat on the Salish Sea to get to the Lummi Reservation from the Olympic Peninsula.

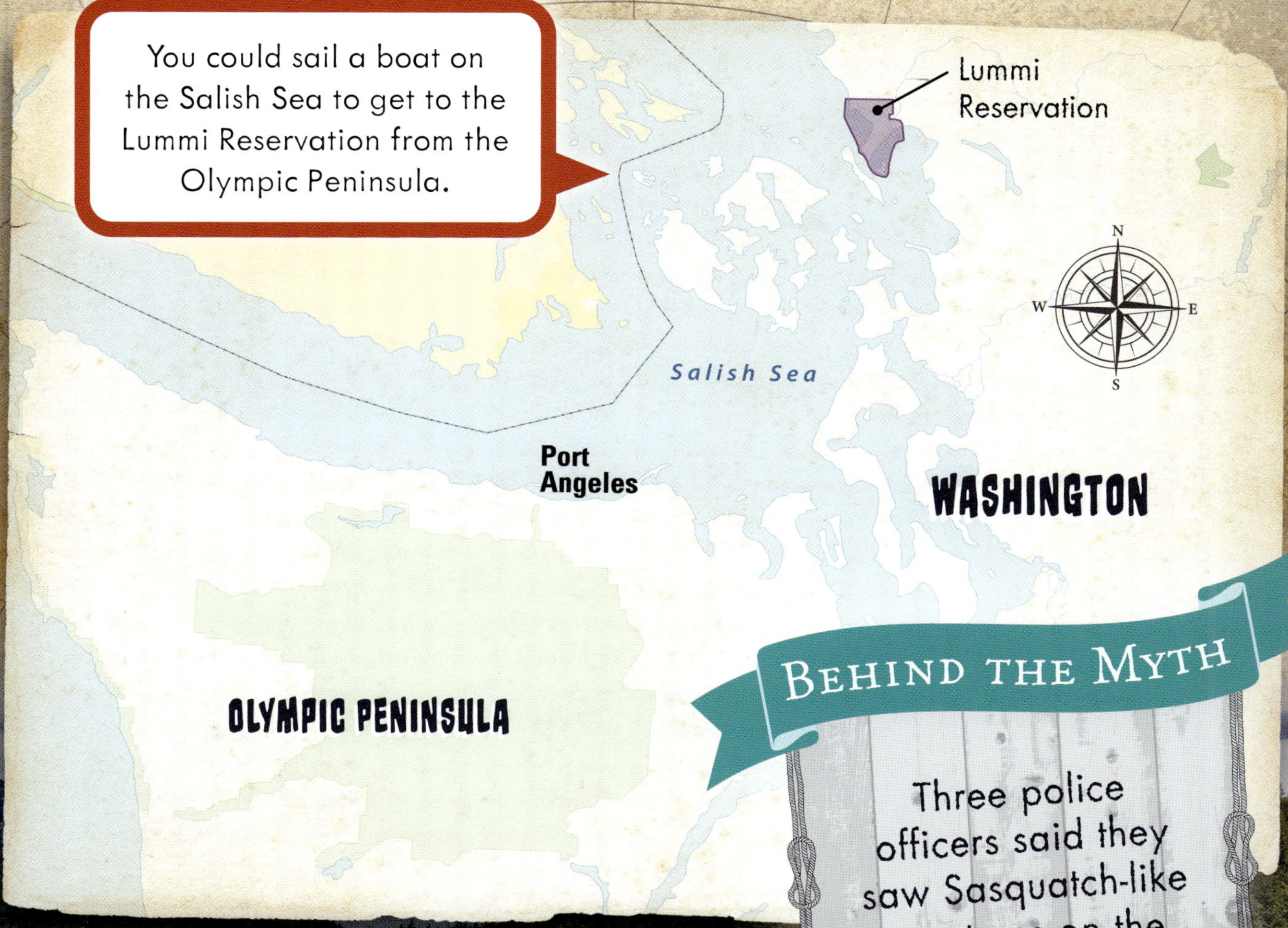

BEHIND THE MYTH

Three police officers said they saw Sasquatch-like creatures on the Lummi Reservation.

The thick forests of Whatcom County could certainly hide a Sasquatch and many other animals. Pictured here is Lummi Island.

BIGFOOT IN BRITISH COLUMBIA

Two famous Sasquatch sightings took place even further north, in British Columbia, Canada. In 1955, William Roe was hiking on Mica Mountain. He said he saw something that looked part-human and part-animal. It was completely covered with brown and silver hair.

A gold prospector, or seeker, named Albert Ostman said a Sasquatch took him to live with its family for a week near Tiba **Inlet** in 1924. Ostman said he escaped when the Sasquatch father became ill.

This map, called a relief map, shows the mountainous parts of British Columbia with brown shading.

Behind the Myth

Some don't believe Ostman's tale because he didn't write about it until 33 years later.

INSIDE APE CANYON

Now head south and cross into the United States. Go to the Cascade Mountains of Washington. In 1924, gold prospectors said they came upon a group of "ape men" near Mount St. Helens. One of the men shot at the creatures, who then ran away.

Later that night, the men were sleeping in a cabin when their house was attacked. They said the ape men threw rocks and beat on the cabin walls. In the morning, they left. The men went back home to Oregon and told their tale.

Mount St. Helens—and Ape Canyon—are in the southwest part of Washington. This mountain is a volcano.

Today, the area of the "ape men" sightings in 1924 is called Ape Canyon. A canyon is a valley that is deep with steep rock sides.

Behind the Myth

The U.S. Forest Service went to the men's cabin in Ape Canyon. They found big rocks and large footprints, but they said the men could have moved the rocks and made the prints themselves.

CAUGHT ON FILM?

Perhaps the most famous Sasquatch sighting happened in California in 1967. Head south from Washington, through Oregon, until you hit Bluff Creek in northern California.

That's where two men, Bob Gimlin and Roger Patterson, were riding horses when they saw a tall, dark, hairy creature with long arms. One of the men managed to record the creature on film before it disappeared. It even turned and looked at them! The men said the creature left footprints. They took pictures of these and made **casts** too.

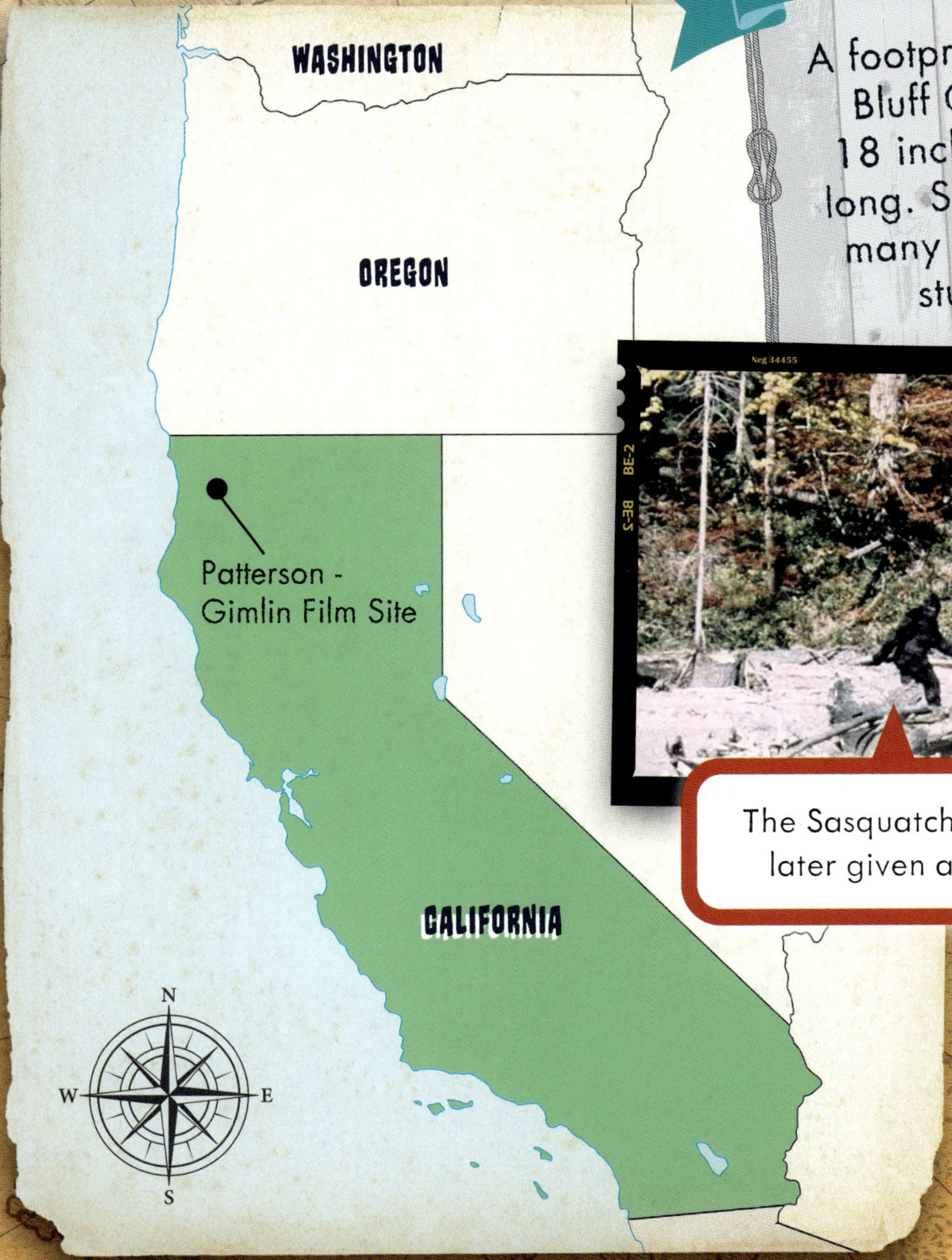

Behind the Myth

A footprint cast from Bluff Creek was 18 inches (46 cm) long. Scientists and many others have studied it.

The Sasquatch in the film was later given a name—Patty.

THE BIGFOOT TRAIL

If there are Sasquatch in California and you want to see one, your best shot might be to hike the Bigfoot Trail. This isn't an official trail, so it isn't as well cared for as a trail taken care of by a park service. That means parts that aren't on roads or paths are hard to hike!

The Bigfoot Trail is about 360 miles (579 km) long. It runs from the Yolla Bolly–Middle Eel Wilderness in the south to the Redwood National and State Parks in the north. One small part takes you through southern Oregon too.

People have counted 32 species, or kinds, of **conifer** trees on the Bigfoot Trail.

Behind the Myth

The U.S. National Park Service calls a wilderness area a place where people can't build on or change the land. People can visit, but they can't stay. What a perfect place for a Bigfoot!

This Bigfoot Trail map has a key, also called a legend. Map keys tell you what colors, lines, shapes, or other things on a map mean. For example, the key tells you the red lines are the path of the trail and the yellow lines are roads.

Colors on this map tell you which areas are wilderness areas and which are national or state parks. The Bigfoot Trail goes through six wilderness areas, which are dark green. It also goes through the Redwood National and State Parks. This area is light green.

Behind the Myth

Some believers think Bigfoot may **migrate** from the mountains to the ocean each year, looking for food!

The Bigfoot Trail was created to help people enjoy nature in northern California. But it's a good place to watch for Bigfoot too!

A redwood may grow to a height of 367 feet (112 m)!

THE BFRO

Hundreds of Bigfoot sightings have happened over the years. A group called the Bigfoot Field Researchers Organization (BFRO) formed in 1995. Its members gather what they call "credible," or believable, Bigfoot stories from around the world. They learn as much as they can about the reports before adding to their collection. They don't accept all reports.

As of October 2024, Washington has the most credible sightings of all U.S. states with 723. California is in second place with 463. British Columbia is the Canadian **province** with the most sightings with 135.

These are the top 10 states for Bigfoot sightings as of October 2024, according to the BFRO website.

Most Bigfoot Sightings, by U.S. State

State	Number of Sightings
Washington	723
California	463
Florida	344
Ohio	326
Illinois	304
Oregon	260
Texas	259
Michigan	225
Missouri	167
Georgia	145

SASQUATCH SKEPTICS

Some people are Sasquatch skeptics. That means they doubt Bigfoot-like creatures exist. They say more and better proof is needed. They think people mistake other animals for Bigfoot or provide fake reports on purpose.

The BRFO has collected more than 5,000 credible reports since the 1990s. These come from every U.S. state except Hawaii. Can they *all* be untrue? Keep your eyes open in the forests, and decide for yourself!

BIGFOOT TRAP

Behind the Myth

A giant ape called *Gigantopithecus blacki* once existed in Southeast Asia. It was 10 feet (3 m) tall! Some think it never died out—and that's what we call Sasquatch.

Some people think Bigfoot sightings may be bears standing on their back feet or some other animal—or people dressed up as Bigfoot!

SASQUATCH MYTHS SPOTTED!

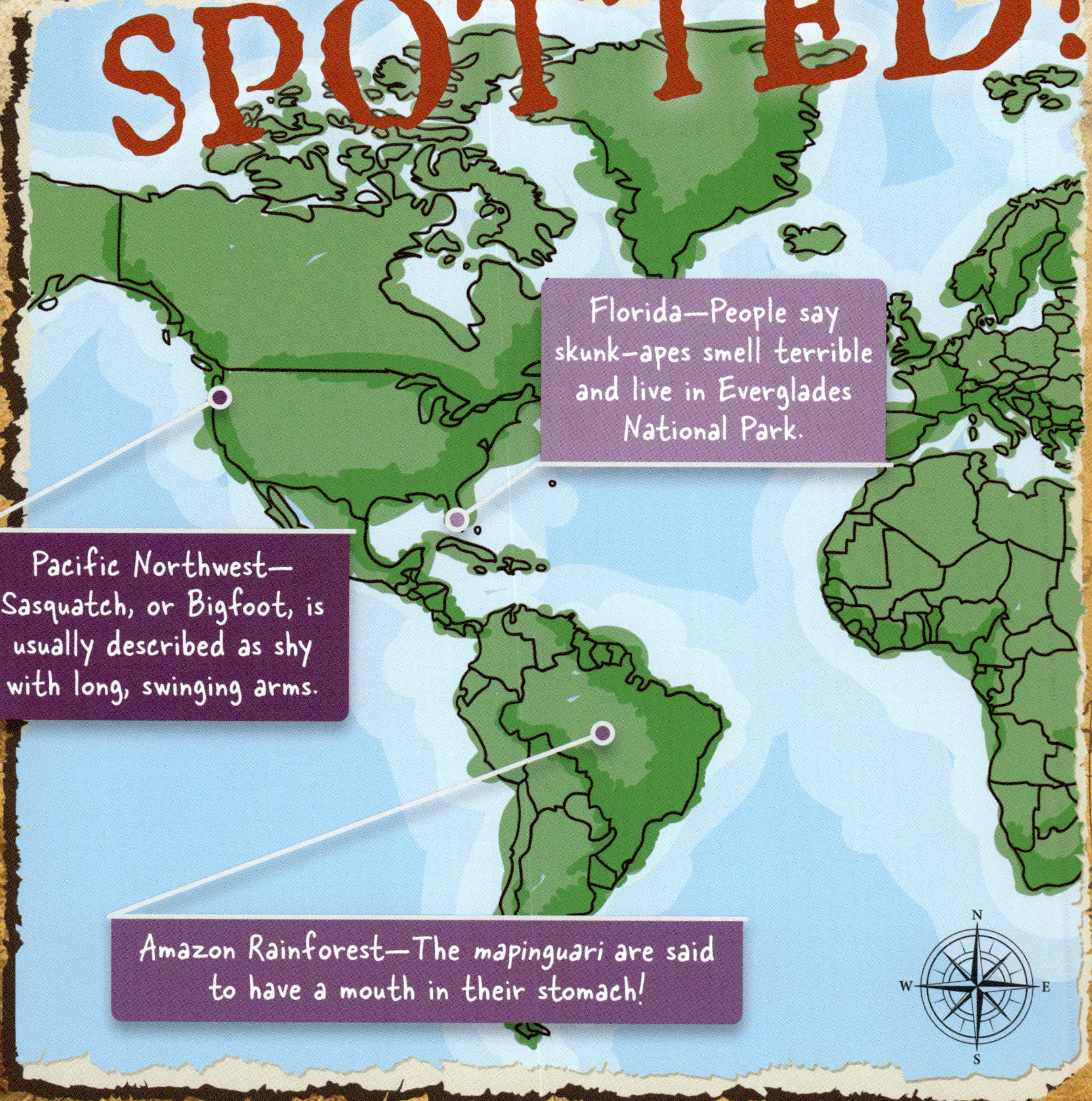

There are many tales of Sasquatch-like creatures around the world. Continue your journey to learn more!

GLOSSARY

cast: An object made by pouring melted material into a mold and letting it harden.

conifer: A tree or bush that often has green leaves all year and that produces cones.

continent: One of Earth's seven great landmasses.

geographic: Having to do with the study of Earth and its landforms.

Indigenous: The earliest known inhabitants of a place.

inlet: A narrow path of water that leads into the land.

mammal: A warm-blooded animal that has a backbone and hair, breathes air, and feeds milk to its young.

migrate: To move from one area to another for feeding or having babies.

peninsula: A narrow piece of land that extends into water from the mainland.

province: An area of a country.

refuge: A place set aside for wild animals to live safely.

reservation: Land set aside by the U.S. government for Native Americans.

volcano: An opening in a planet's surface through which hot, liquid rock sometimes flows.

FOR MORE INFORMATION

Books

Quick, Megan. *Do You Dare Visit the Bigfoot Trail?* Buffalo, NY: Enslow Publishing, 2024.

Stickney, Laura. *Sasquatch*. Minneapolis, MN: Kids Core, 2024.

Websites

Bigfoot
kids.britannica.com/students/article/Bigfoot/631810
Check out this article about Bigfoot descriptions.

Bigfoot Evidence
www.history.com/videos/bigfoot-evidence
Watch the Patterson-Gimlin film and decide if it's real!

Publisher's note to educators and parents: Our editors have carefully reviewed these websites to ensure that they are suitable for students. Many websites change frequently, however, and we cannot guarantee that a site's future contents will continue to meet our high standards of quality and educational value. Be advised that students should be closely supervised whenever they access the internet.

INDEX